ANIMAL RHYME TIME!

RHYME TIME WITH CROCODILES!

BY LEIGH MCCLURE

Gareth Stevens PUBLISHING

Please visit our website, www.garethstevens.com. For a free color catalog of all our high-quality books, call toll free 1-800-542-2595 or fax 1-877-542-2596.

Library of Congress Cataloging-in-Publication Data

Names: McClure, Leigh, author.
Title: Rhyme time with crocodiles! / Leigh McClure.
Description: Buffalo, NY : Gareth Stevens, [2024] | Series: Animal rhyme time! | Includes index.
Identifiers: LCCN 2022048283 (print) | LCCN 2022048284 (ebook) | ISBN 9781538285305 (library binding) | ISBN 9781538285299 (paperback) | ISBN 9781538285312 (ebook)
Subjects: LCSH: Crocodiles–Juvenile literature.
Classification: LCC QL666.C925 M395 2024 (print) | LCC QL666.C925 (ebook) | DDC 597.98/2–dc23/eng/20221031
LC record available at https://lccn.loc.gov/2022048283
LC ebook record available at https://lccn.loc.gov/2022048284

Published in 2024 by
Gareth Stevens Publishing
2544 Clinton Street
Buffalo, NY 14224

Designer: Claire Wrazin
Editor: Caitie McAneney

Photo credits: Cover, p. 1 diegooscar01/Shutterstock.com; p. 5 Uwe Bergwitz/Shutterstock.com; p. 7 Marina Demkina/Shutterstock.com; p. 9 Isabelle van Mierlo/Shutterstock.com; p. 11 Arunee Rodloy/Shutterstock.com; p. 13 Artush/Shutterstock.com; p. 15 Nelia L/Shutterstock.com; p. 17 Willyam Bradberry/Shutterstock.com; p. 19 JMx Images/Shutterstock.com; p. 21 Jesus Cobaleda.

Printed in the United States of America

CPSIA compliance information: Batch #CS24GS: For further information contact Gareth Stevens, New York, New York at 1-800-542-2595.

CONTENTS

Boldface words appear in the glossary.

Cool Crocodiles!

What's that thing swimming in the bay?

It's a crocodile—run away!

Crocodiles are strong and cool,

but you don't want one in your pool.

These **fierce** hunters have toothy smiles.

Saltwater crocs are the largest **reptiles**!

SALTWATER CROCODILE

What's a Crocodile?

Part of the Crocodilia order of reptile, there are many kinds of crocodiles. Crocs are cold-blooded and lay eggs. They have thick skin and short legs. Crocodiles have a V-shaped **snout**. They lived on when dinosaurs died out!

Big Crocs!

Saltwater crocs can grow 23 feet in length.

They are known for their great strength!

At weights of over 2,200 pounds,

they have the strongest bite around.

The smallest is the African dwarf crocodile.

At 6 feet long, it's still one big reptile!

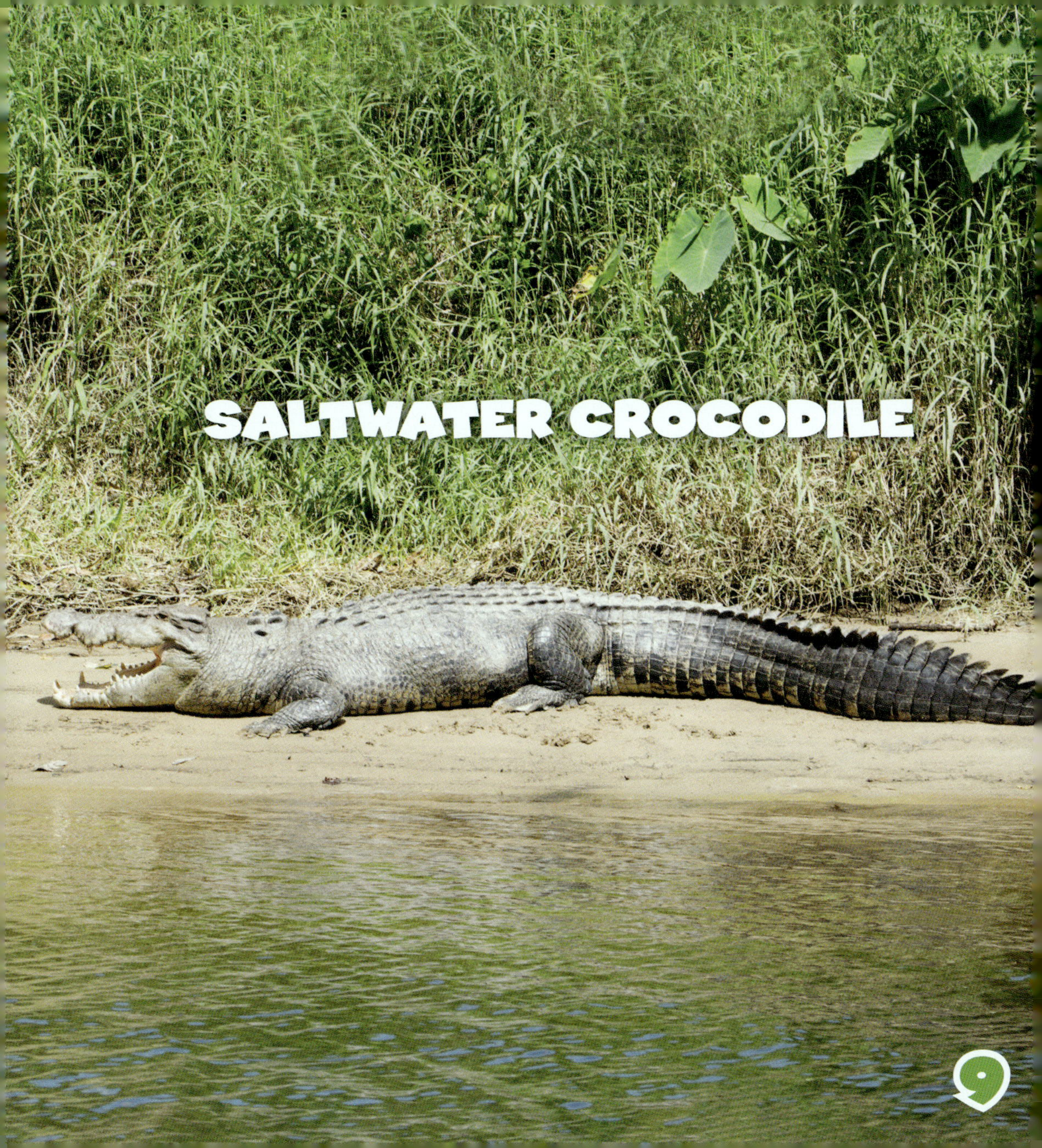
SALTWATER CROCODILE

Long Lives

Baby crocs **hatch**, helpless and small.
Moms take them to water, caring for all.
Male crocs fight off anyone close.
As a crocodile ages, it grows.
With strong eyesight and sharp ears,
wild crocs live up to 70 years!

Where on Earth?

Crocodiles live in places that are hot.

In Africa and Australia, you'll find a lot!

They live in the Americas and Asia too.

The island of New Guinea even has a few!

Many are found swimming in the Nile.

India is home to the mugger crocodile!

NILE CROCODILE, NILE RIVER

In which **habitats** can you find a croc?
Spot them catching sun on a rock.
Crocs live in salt water, fresh water too.
Lakes and rivers are home to this crew.
Mother crocs make nests on the shore.
Crocs can swim to sea for food **galore**.

Terrific Teeth!

It's clear to see why crocs are in charge:

crocodile teeth are pointy and large.

They cut through **prey** like a knife.

Crocs lose many teeth in life.

But with teeth, crocs are never poor.

They can grow around 8,000 more!

On the Hunt

What do crocodiles like to eat?

They're carnivores, so they eat meat.

Crocs hunt animals big and small.

From fish to buffalo—they'll eat it all!

Deer and pigs are favorite picks.

Ambushing is their special trick!

Crocs on Top

One more fact before we stop!

On the **food chain**, crocs are at the top.

Crocs are as fierce as animals get.

But humans are their biggest **threat**.

Alive since the time of the dinosaur,

may they live for many years more!

GLOSSARY

ambush: To attack from a hiding place.

fierce: Wild, strong, and likely to fight.

food chain: A series of living things in which one uses the next lowest one as food.

galore: In large amounts.

habitat: The natural place where an animal or plant lives.

hatch: To break open or come out of.

prey: An animal hunted by other animals for food.

reptile: An animal covered with scales or plates that breathes air, has a backbone, and lays eggs, such as a turtle, snake, lizard, or crocodile.

snout: An animal's nose and mouth.

threat: Something likely to cause harm.

FOR MORE INFORMATION

BOOKS

Katz, Susan B. *Is It an Alligator or a Crocodile?* North Mankato, Minnesota: Pebble Books, 2022.

Levy, Janey. *Nile Crocodiles Bite!* New York, NY: Gareth Stevens Publishing, 2021.

WEBSITES

Alligators and Crocodiles
www.ducksters.com/animals/alligatorcrocodile.php
What's the difference between an alligator and a crocodile? Find out here!

Nile Crocodile
kids.nationalgeographic.com/animals/reptiles/facts/nile-crocodile
Learn more about the mighty Nile crocodile with National Geographic Kids.

INDEX